Dark Matter Discovered

by Rolf A. F. Witzsche

Contents

3

About the Illustrated Science series
On the Ice Age and Climate Change
and the book

Dark Matter Discovered

Dark matter is theorized to exist in order to fulfill certain postulates that are fundamentally erroneous. However, invisible forms of high concentrations of mass do exist, though in a different context than the one theorized. Are you confused yet?

The fact is; plasma in space, while is electrically charged, also has mass. All the mass in the universe is located in the protons and electrons that atoms were created from. Atoms have a low mass, because their size is typically 100,000 times larger than the parts they are made of. But plasma is free flowing. It can be dynamically compressed, electromagnetically, to extremely high-density formation. The densest mass-concentrations in the universe are of this type. All the anomalies that have been observed, are simply plasma structures, with features that can be replicated in the small in the laboratory. No magic is needed. This does not mean that the dynamics are simple.

Plasma in the physical universe is as challenging in perception as the spiritual domain in the human sphere. Both are invisible, except by their effects, but they are understandable and knowable.

On the same basis it becomes possible to recognize the outcome of plasma physics operating in the universe. This is important in astrophysics, because of the huge consequences that follow, when we mess up in our thinking.

With the Ice Age Challenge now before us, we face two imperatives. One is to understand the physical dynamics, and to create the physical infrastructures that enable human living to continue in an Ice Age climate. The second challenge, the greater challenge, is to raise up our humanity to such height as will impel us to get the job done. Some say that miracles are needed on both fronts. But what of it? Are we, as human beings, not the miracle makers on the Earth?

In the real universe, the cosmic operations are anti-entropic in nature, and expanding and progressing. We, ourselves are evidence of this progression. Should this progression have ended? Neither is our Sun isolated from the progressive nature of the universe, but expresses its dynamics, its resonating plasma streams, and their reflection in the climate on Earth. Shouldn't we develop ourselves spiritually and culturally, likewise?

Climate Change reflects the nature of the universe. It should also be reflected in us.

The Earth itself is the creation of the Sun, with its atoms having been massively synthesized in high-energy times near the center of the galaxy.

The synthesizing plasma fusion is presently at a low state, though it is currently enhanced for our Sun by electromagnetic 'Primer Fields' that focus interstellar plasma onto the Sun in a highly condensed manner. When the plasma-focusing system becomes inactive, below the required threshold conditions, the Sun reverts to a type of cosmic default level with 70% less energy being radiated, and higher rates of solar cosmic-ray flux being experienced.

At the present rate of plasma diminishment being experienced, the solar activity phase-shift threshold to the next Ice Age period may be crossed in 30 years, or in the 2050s, most likely. With the primer-fields system gone inactive by then, the climate on Earth will get 40 times colder than the Little Ice Age in the 1600s had been. Ice core evidence promises that. Without the needed preparations for human living in such an environment, 99% of humanity would die of starvation, both by the cold, and by CO_2 depletion that diminishes agriculture, as more CO_2 becomes dissolved into the sea.

With the 'Primer Fields' being critical for our very existence, the exploration of them is likewise critical.

In the Little Ice Age, between 10% and up to 30% of the populations in Europe had perished by starvation. The last Big Ice Age was evidently vastly harsher. Only 1-10 million people emerged from it alive. That's all we had after 2 million years of development. We want to do far better this time around; and we can, with large-scale technological infrastructures for our food supply. But will we create them? Will we get the job done in the 30 years that we still have left before the Ice Age starts anew? Will we even consider it? And how certain are we that the phase shift to the next glaciation period will begin, as the evidence suggests, in the 2050s? We have no slack on this front. Should we fail us on this absolute front, we would be committing suicide.

Numerous fields of evidence tell us that the next Ice Age is near. That's where the truth begins. Most of the evidence was discovered in the 1990s and thereafter. Some evidence is measured in ice cores; some is measured in space, by satellites. Some measurements are also made on the ground in terms of measurements of the Earth's magnetic-pole drift observed in northern Canada. All of this is seen combined with high-energy physics experiments at a leading national laboratory, and is also explored in the small in static experiments.

So, what will the answer be? Will we move with the evidence? Or will we lay ourselves down to die by default?

It takes an independent researcher to brake the taboos that have kept mainstream cosmology imprisoned, increasingly, during the past century, even while what is regarded as taboo is known to be wrong.

The Illustrated Science series is intended to open the scene beyond the threshold of accepted taboos, to where the actual physical evidence speaks for itself.

The scope of the existential challenge that the Ice Age brings with it, takes astrophysics out of the academic domain and places it into the foreground as one of the most-critical issues of our time. The big Climate Change events that have already worldwide effects are mere fringe effects in the flow of the ever-changing cosmic dynamics. The big effect, when the Ice Age begins anew, promises to be caused by a dimmer and colder Sun. The loss of 70% of the Sun's radiated energy defines our climate future that begins in the near term.

Sure, we can live with all that by creating new platforms for agriculture that are able to operate under Ice Age conditions. But will we do it? The task is enormous. Or will we fail ourselves on this front? We have no reason to allow us to fail. We have the materials and energy resources on hand to accomplish everything that is required for us to continue to live in an Ice Age World. But will we do it? The big question that never goes away, therefore, is; will we develop our inner resources as human beings sufficiently to get the job done, and to get it done in time? Or will we do nothing, ignore the challenge, and condemn our children and one-another to an agonizing death by starvation? That's the choice.

Towards meeting the inner challenge, I have created the epic series of novels, The Lodging for the Rose. And further, towards meeting the science challenge, I have produced numerous research books and several dozen exploration videos that the Illustrated Science series is modeled after. The work is the result of a quarter century of research, for which numerous elements of evidence in related fields came to light during the timeframe of my research.

It is my hope that the work that went into all of these projects will help in some degree - for humanity that we are all a part of - to write itself a ticket to have a future.

High-resolution color images, of the images in this book, can be obtained at www.iceagetheatre.ca

Does Dark Matter exist?

NASA - galaxy cluster (CL0024+17)

Dark matter is a hypothetical phenomenon. It is a type of matter that cannot be seen with telescopes.
Is that a paradox: Matter that cannot be seen?
It is deemed to exist in such abundance that it accounts for most of the matter in the Universe.
Does the paradox deepen?
The existence of dark matter is inferred from its effects on the visible universe. While its presence cannot be detected directly, its effect reveals its existence. This makes dark matter one of the greatest mysteries in modern astrophysics. Some say, that up to 90% of all the matter of the universe exists as "dark matter," or a subset of it that is termed "dark energy."
This doesn't solve the paradox, does it?
Actually, the existence of dark matter was theorized as a means for solving a great paradox in astrophysics.

Postulated to account for the stark discrepancies

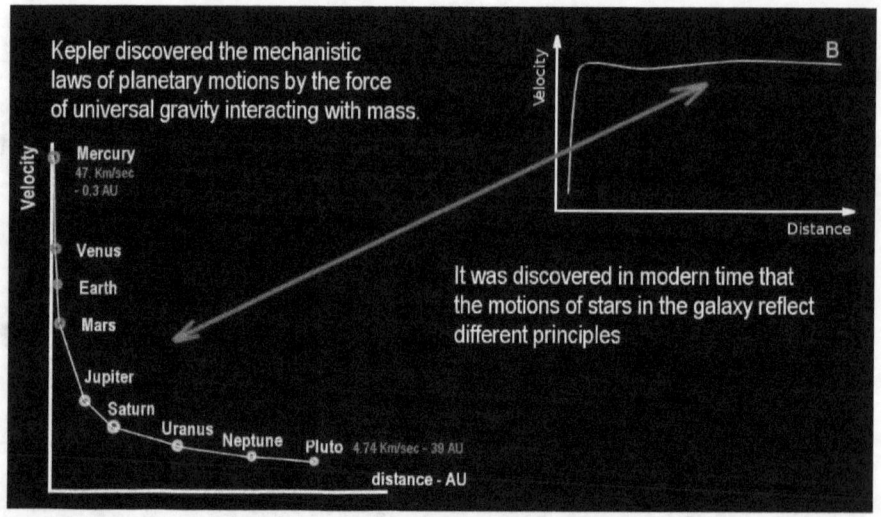

The existence of dark matter was first postulated in the early 1930s to account for the stark discrepancies between observed 'orbital' velocities of stars in the Milky Way and the laws of orbital mechanics discovered by Johannes Kepler in the early 1600s.
In our solar system the orbital velocity of planets diminishes with the square of the distance. The velocity of the planets at the lower left reflects the gravity forced dynamics of orbital motion that Johannes Kepler had discovered. With the force of gravity diminishing with the square of the distance, the outer planets move more slowly, because if they didn't, the centrifugal force of their movement would drive them out of the solar system.
Ironically, the opposite relationship between speed and distance has been measured in the motion of stars. It has been reasoned from this observation that the mass distribution in a galaxy must be vastly different than it is in a solar system, because stellar orbits are simply not possible under the laws that Kepler had dealt with.

Large invisible mass that somehow solves the paradox

Stellar orbits arround the galactic center are NOT possible under Keper's laws of gravitational mechanics. However, no other perception is allowed. Thus new mythical 'epicycles' are imagined to make the modern doctrine plausible.

In order for this thing to work, it has been postulated that the entire galaxy must be surrounded and be pervaded by a large invisible mass that somehow solves the paradox. But that's like saying, "we really don't know how this thing works."

The problem is, that this approach, such as adding fudge factors and imaginary entities which can't actually be detected, doesn't really work, does it? It all adds up to grand fairy tales that one is supposed to take on faith. But where do we find the truth by which everything makes sense, and which can be recognized to be truthful?

To answer this question, we need to take a step back and look at what the paradoxical problem is based on.

11

The answer is that our box is too small

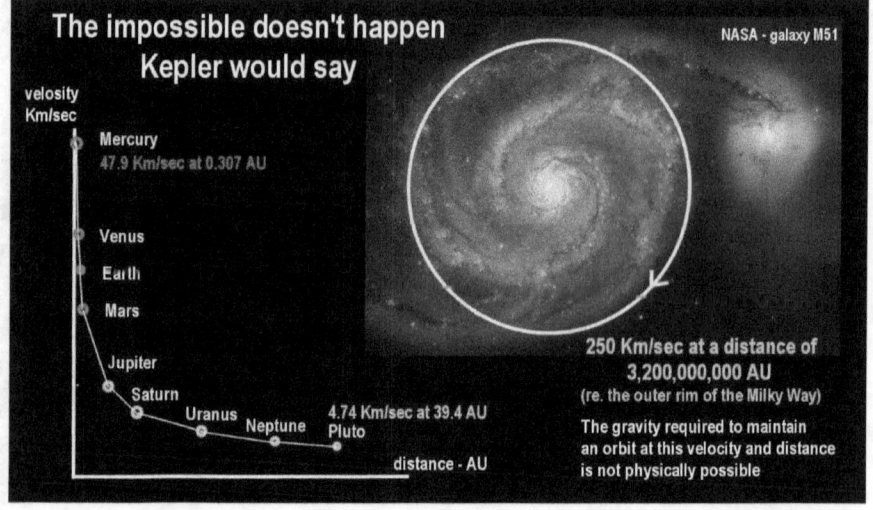

The impossible doesn't happen
Kepler would say

NASA - galaxy M51

velosity
Km/sec

Mercury
47.9 Km/sec at 0.307 AU

Venus

Earth

Mars

Jupiter

Saturn

Uranus

Neptune

4.74 Km/sec at 39.4 AU
Pluto

distance - AU

250 Km/sec at a distance of
3,200,000,000 AU
(re. the outer rim of the Milky Way)

The gravity required to maintain
an orbit at this velocity and distance
is not physically possible

When one looks at the paradoxical problem from a higher-level
standpoint where gravity is not the only force in the universe,
suddenly an answer becomes fairly obvious.

The answer is that our box is too small in which the paradoxical
reasoning is conducted. The entire problem of paradoxical stellar
movements is based on the assumption that only the weakest force
that exists in the universe, the force of gravity, may be considered,
which in addition has an effect that diminishes with the square of
the distance from its source.

While we see these factors being dominant in the intimate domain
of the orbits of planets around a sun, where the planets are close
enough to be affected by gravity, different principles evidently
come into play at the larger distances where gravity no longer has a
hold on celestial objects, where higher-order principles affect the
movements of the stars.

The effect of any central gravity is zero

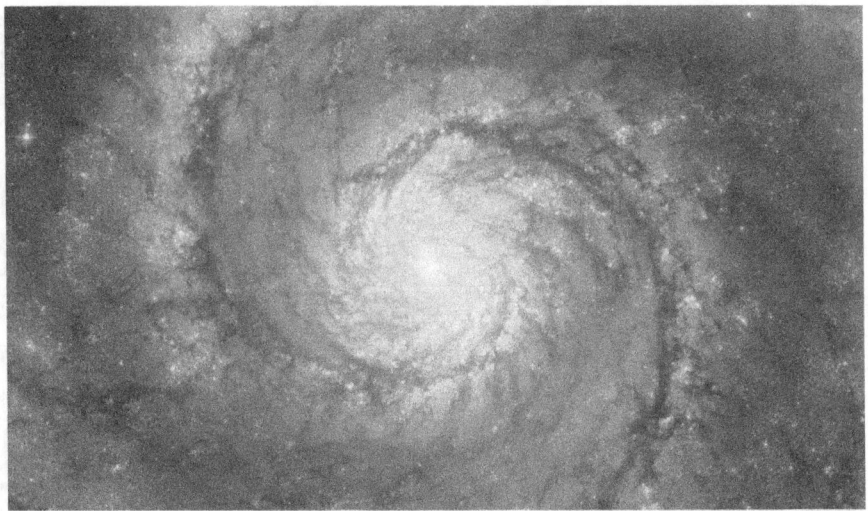

When we deal with distances in the order of tens of thousands of light years, the effect of any central gravity is zero, no matter how massive the central gravity might be.

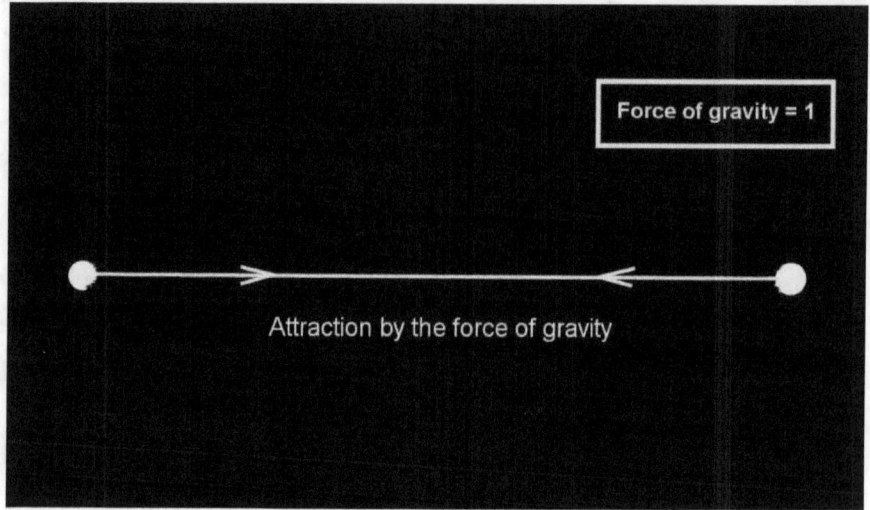

Force of gravity = 1

Attraction by the force of gravity

But why would we limit ourselves to considering only the force of gravity, which is after all the weakest force in the universe, which in addition to being the weakest force, also diminishes with the square of the distance. Why would we keep our box of perception limited to only that? Why do we keep it that small?

The electric force is deemed not to exist in cosmic space

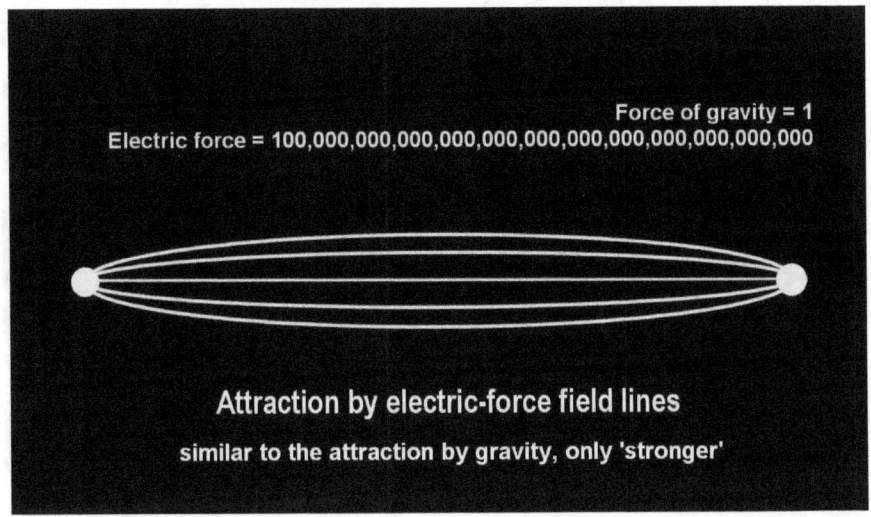

It is a well-known fact in physics that a vastly stronger force than gravity exists, which in addition has an infinite reach. This vastly greater force, is the electromagnetic force. The electric force of attraction between two charged particles, is by far the strongest long-reaching force in the universe. It is 39 orders of magnitude stronger than the force of gravity. Nor does it diminish with the square of the distance as gravity does.

The problem is, that the electric force is deemed not to exist in cosmic space. By some high-ranking command of the keeper of the box, the electric force is not allowed to be considered. The command is, that it does not exist.

Plasma is rare on earth, but it is native to space

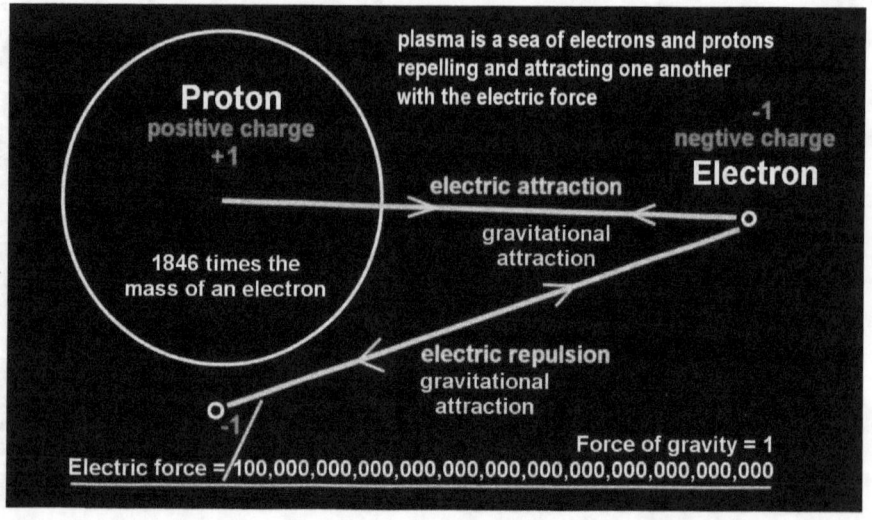

Of course it does exist. It exists in the form of plasma in space.
Plasma is rare on earth, but it is native to space. Plasma is a soup
made up of the two basic particles that all atoms in the universe are
made of. They are named protons and electrons. Both carry a
complimentary electric charge. The proton has a positive charge,
and the electron a negative charge. Both particles also have a
specific size and mass. The proton is 1800 times more massive. By
their mass, the plasma particles have a gravitational attraction to
each other and to all forms of mass. But by their electric interaction,
as I said before, the acting force is 39 orders of magnitude
stronger.

However, and here it gets interesting, both of these plasma
particles are extremely small. They are too small to ever to be
visible.

The largest of the two, the proton, is 100,000 times smaller than
the smallest atom in the universe, and the electron is a thousand
times smaller than that. These particles are simply invisible. Their
invisibility qualifies them for the term, "dark matter." Nobody can

see plasma. Nor does plasma ever emit light. Plasma is dark, transparent, and invisible, even while it is the primary state of all mass in the universe.

That 99.999% of the mass of the universe exists in plasma form

Researchers at the Los Alamos National Laboratory have come to the conclusion that 99.999% of the mass of the universe exists in plasma form and less than a thousandth of a percent exists in atomic form. In other words, the universe exists primarily of vast volumes of dark matter that carries an immense electric charge.

Only atomic structures emit light

The dark matter becomes 'visible' only when the movement of plasma particles affects the electron movement in the shells of atomic elements. The resulting agitation causes the atoms to emit light. Only atomic structures are able to emit light. When we see light in space, we see atoms agitated by dark matter in motion. The dark matter itself remains dark at all times.

Every sun in the universe is lit up by this process

NASA - view from ISS

Every sun in the universe is lit up by this process of dark matter in motion interacting with atomic elements. In the case of the Sun, the plasma particles, the fabled particles of dark matter, become fused into atomic elements. Where this fusion takes place, the interaction is extremely intense, and so is the resulting light. The emitted light is energy derived from moving plasma. This is also where the critical function of dark matter is located.

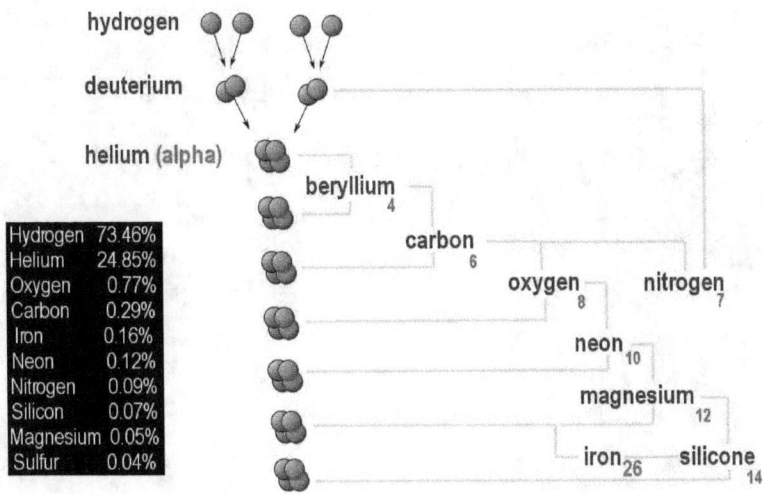

Dark matter is not a dead thing that hangs in space like a fog. It is electric energy in motion, and the motion itself is caused by the fusion process on the surface of a sun. When the electrically charged plasma particles, also termed dark matter, become fused together into atomic elements that are electrically neutral, a portion of the electric force becomes neutralized. It ceases to exist.

The plasma-fusion process thereby becomes a sink

The faucet as a sink

The plasma-fusion process thereby becomes a sink in which dark matter is 'consumed' and turned into atomic elements that flow away with the solar wind. The consumption keeps the plasma streams flowing.

Plasma streams are electricity in motion

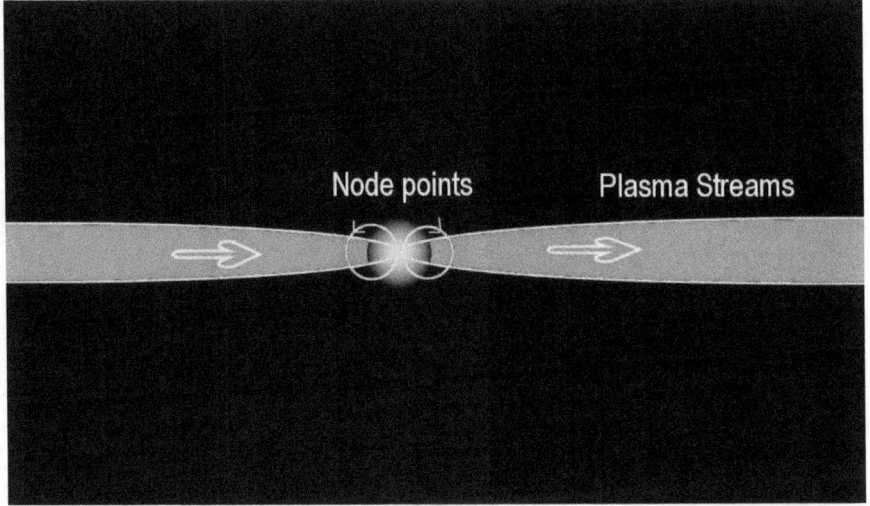

The flowing plasma streams are electricity in motion. When electricity flows in two parallel wires in the same direction, the two wires are drawn together by the magnetic field that the movement of electricity creates. When electricity moves in space as plasma, the entire dark-matter soup becomes pinched together ever tighter, which of course increases the magnetic field and the pinching effect. Eventually a point is reached where the plasma gets all tangled up and a type of explosion occurs.

What happens when the pinch effect goes critical?

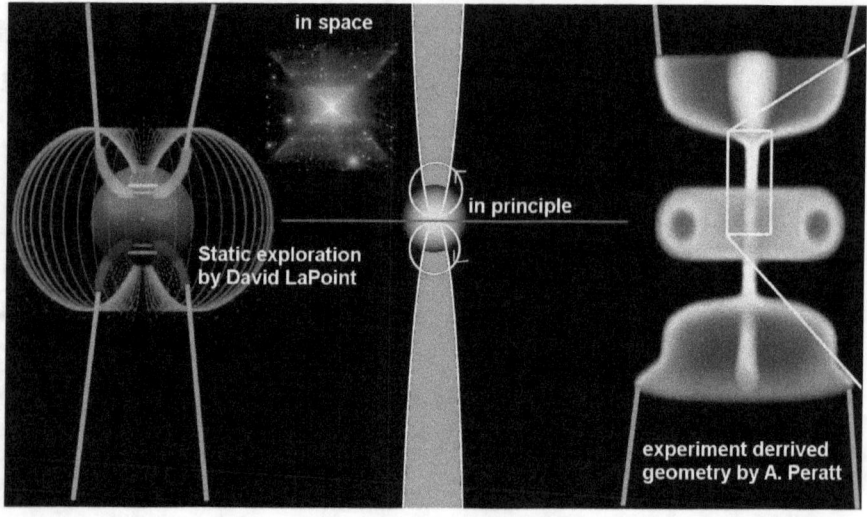

A number of research efforts have been made to explore what happens when the pinch effect goes critical. The blue geometry was produced in a high-power experiment at the Los Alamos National Laboratory. It was discovered that at the tightest pinch, the plasma gets magnetically curled backwards into a magnetically contained bulb where the pressure builds up. Another researcher explored this region statically, to explore how the process works. He termed the magnetic fields, the primer fields.

In the dynamic experiment, the excess energy in the system that caused the explosion at the node point gets dissipated in a hollow ring-type structure. In space, these structures are rarely visible, as dark matter simply isn't visible.

NASA has photographed two gigantic structures

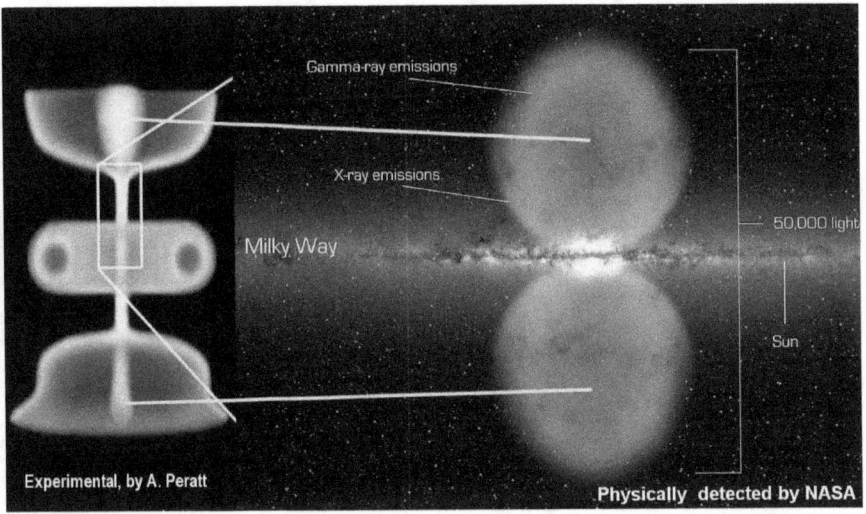

Only in extreme cases can we actually see some parts of them. One such case where parts of the primer fields become visible, is our own galaxy. NASA has photographed these two gigantic structures of concentrated dark matter, shown here, which are visible by their interacting with faint atomic elements, thereby emitting x-ray light, and gamma-ray light.

Large primer fields too faint to be detectable

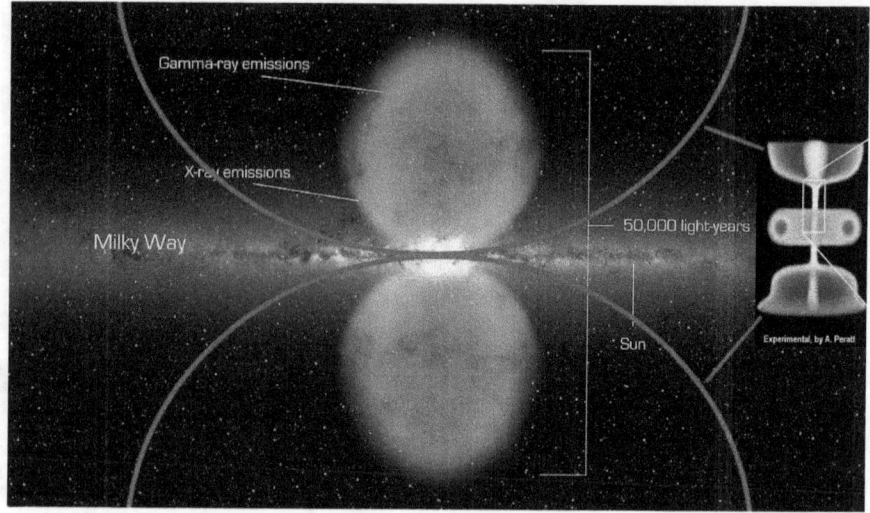

The large primer fields, however, are evidently too faint to be detectable with available instrumentation. However, what we see here, and also what we cannot see directly of the dynamic movements of dark matter, solves the original paradox of the movements of the stars in our galaxy.

Stars moved by electromagnetic dynamics

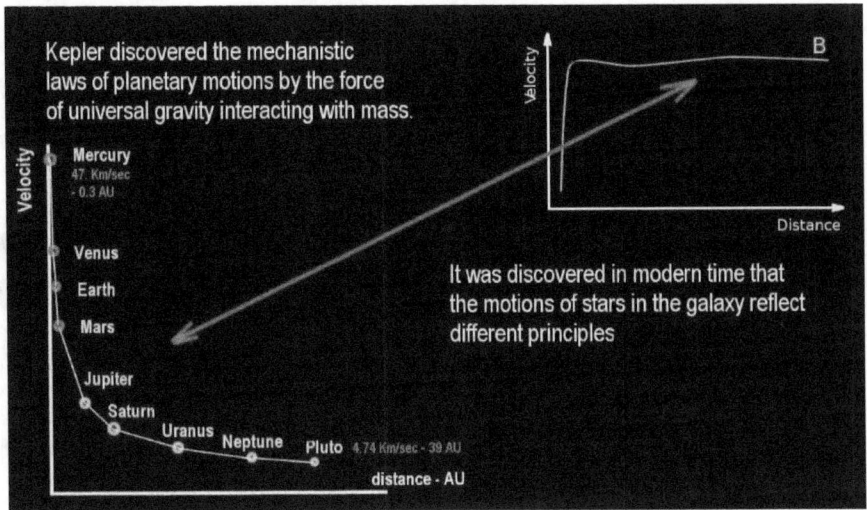

Kepler discovered the mechanistic laws of planetary motions by the force of universal gravity interacting with mass.

Mercury 47. Km/sec - 0.3 AU

Venus

Earth

Mars

Jupiter

Saturn

Uranus

Neptune

Pluto 4.74 Km/sec - 39 AU

distance - AU

It was discovered in modern time that the motions of stars in the galaxy reflect different principles

The stars are not orbiting in a gravitationally bound system, but are moved by the electromagnetic dynamics of the primer fields system that is motivated by the plasma streams of dark matter.

27

This means that the entire galaxy is rotating

This means that the entire galaxy is rotating as a complete entity while it is constantly expanding from within.

Collectively powers entire clusters of galaxies

NASA - galaxy cluster (CL0024+17)

Evidence also exists that the same process that powers galaxies, collectively powers entire clusters of galaxies. The blue ring-image that you see here, that is centered on a cluster of large galaxies, is obviously not the result of gravitational lensing, since gravitational lensing doesn't affect light in this manner. Instead, the blue image that you see reflects in the large the typical geometry that has been discovered in the laboratory in high-energy experiments.

Plasma combines into a ring of 56 filaments

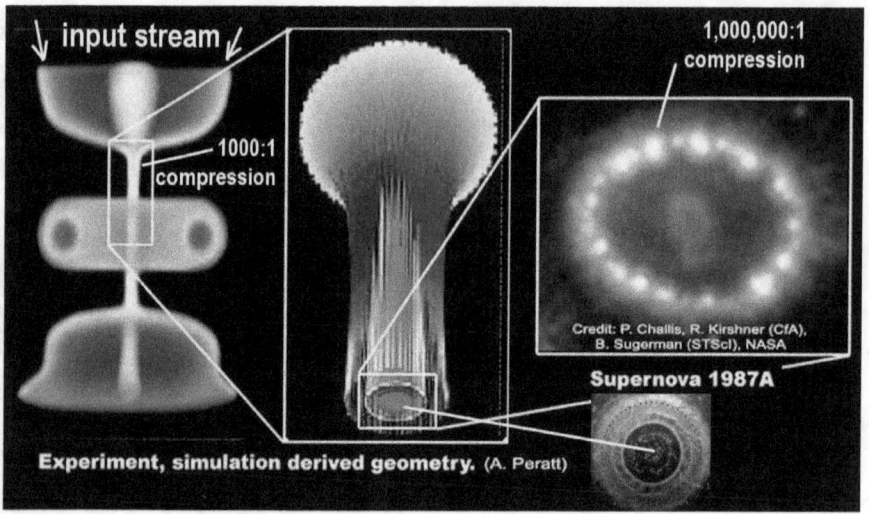

It has been discovered in the lab that by extreme concentration, plasma combines into a ring of 56 filaments that eventually combine into 28 filaments, then 14, and so on. One such ring of filaments has been discovered centered on supernova 1987A.

*The ring structure

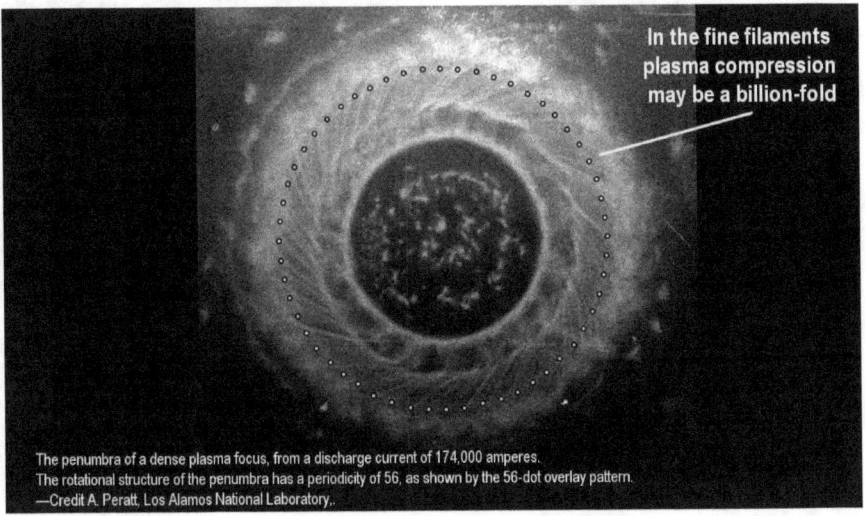

In the fine filaments plasma compression may be a billion-fold

The penumbra of a dense plasma focus, from a discharge current of 174,000 amperes.
The rotational structure of the penumbra has a periodicity of 56, as shown by the 56-dot overlay pattern.
—Credit A. Peratt, Los Alamos National Laboratory,.

The ring structure shown here is typical for high-energy plasma streams.

Looking down the 'barrel' of a massive plasma stream

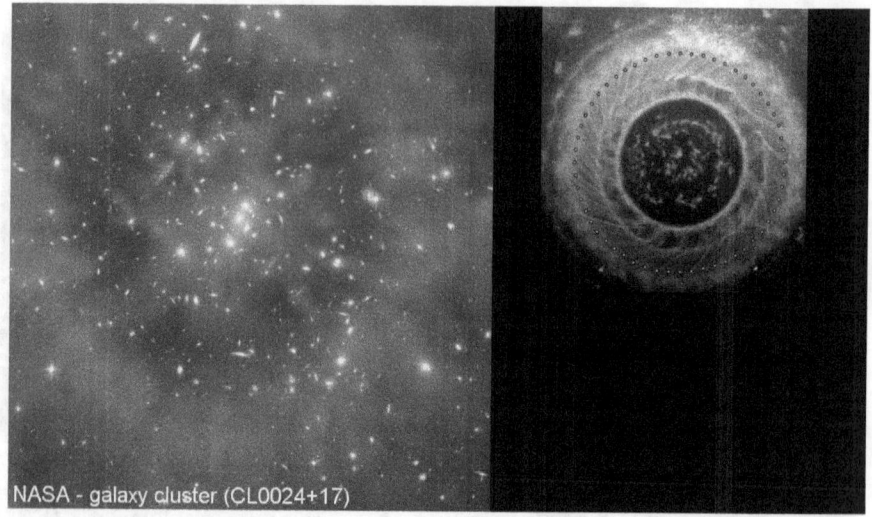

NASA - galaxy cluster (CL0024+17)

This type of pattern is apparent in the blue ring that is centered on a large galaxy cluster. The galaxy cluster doesn't create the ring by bending its light. The ring is formed by the immense plasma streams that are feeding the cluster. We appear to be looking down the 'barrel' of a massive plasma stream that is feeding its black matter into the galaxy-cluster system, which powers the cluster. The blue light of the ring is emitted by atomic elements in the path of the plasma stream.

The circular light show

Plasma lensing

Hubble Space Telescope in Abell 1689

The circular light show also appears faintly when no atomic elements are encountered in the path. In this image we look once again down the axial center of a giant plasma stream system, streaming towards a major galaxy cluster. The faint arcs appear to be light reflected by the different inner ring structures in the plasma stream.

Their effect on the alignment of planets and galaxies

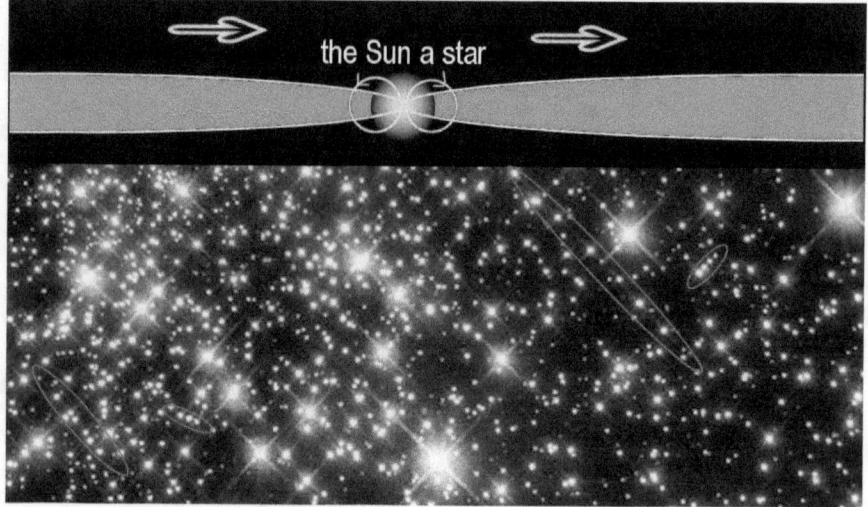

Another way by which we can see the existence of dark matter or plasma streams in space, is by their effect on the alignment of planets and galaxies. While a plasma fusion on a sun motivates the plasma flow, the sun consumes only a part of the plasma, the rest flows on to the next star in line. The electric attraction that is propagated through the plasma streams, ties the connected stars into linear clusters like beads on a string. While we look at three dimensional space flattened out in this two dimensional image, the linear alignment pattern that often results, and is visible in star fields, is a clearly visible feature wherever one finds dense star formations.

The same linear pattern

strings of galaxies and stars

ESO/VIMOS galaxy cluster ACO 3341

ESA/Hubble & NASA
Acknowledgement: Claude Cornen

The same linear pattern that we see in fields of stars is also visible in fields of galaxies.

The Capodemonte Deep Field

The Capodimonte Deep Field - of 35,000 galaxies

NASA

In this partial field of 35,000 galaxies, termed the Capodemonte Deep Field, many strings of galaxies are visible, including some rather long strings. And it is here, where the existence and characteristics of dark matter in space is critically important for us on the Earth. Our galaxy, the Milky Way Galaxy, is just a link in one of these types of chains of galaxies.

Each has a different electric resonance characteristic.

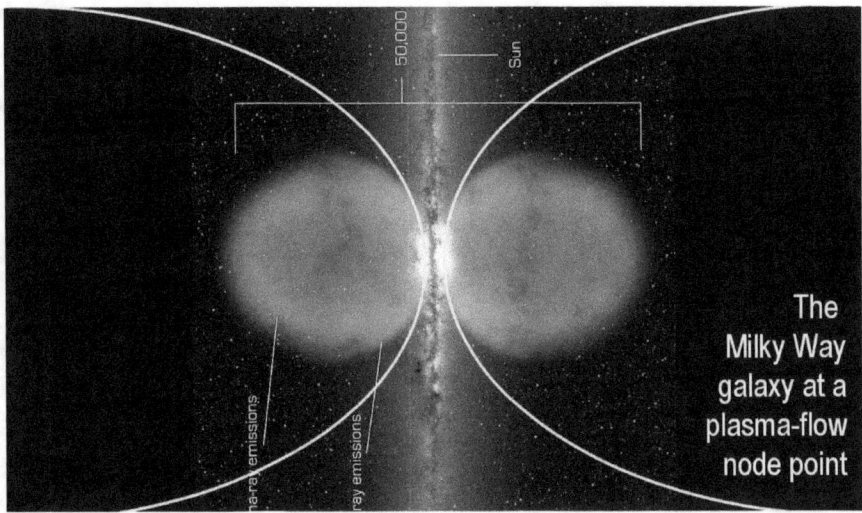

Our galaxy has an in-flowing plasma stream and an out flowing plasma stream. Depending on the length of these streams, each of them has a different electric resonance characteristic.

From their combination the great Ice Age epochs resulted

Galaxies at node points

In our case, one of the streams has a 140-million-year resonance, approximately, and the other a 62-million-year resonance. Both of them, together, reflect themselves across the entire galaxy, affecting the plasma density throughout time, which in turn affects the climate on Earth in a big way. From their combination the great Ice Age epochs resulted.

When the plasma density is high, the Sun is more active

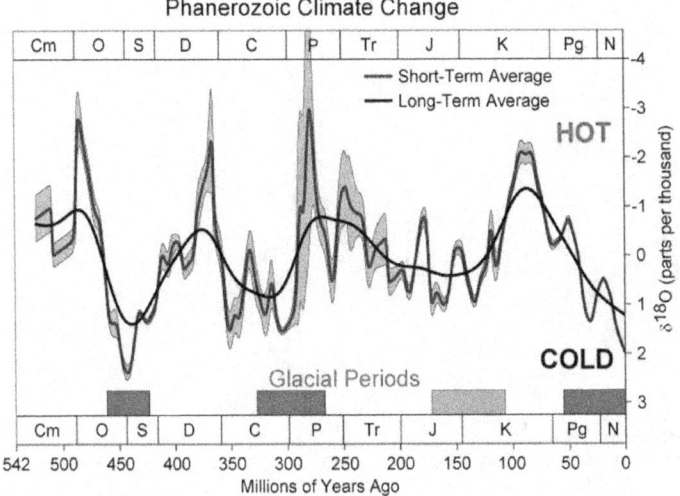

Phanerozoic Climate Change

When the plasma density is high, the Sun is more active, and the Earth gets warmer. These up and down pattern reflect the combination of the 140 and 62-million-year resonance cycles, reflected in historic temperature change gleamed from deep sea sediments by analyzing the changing oxygen-18 isotope ratios.

We see two deep Ice Age epochs occurring

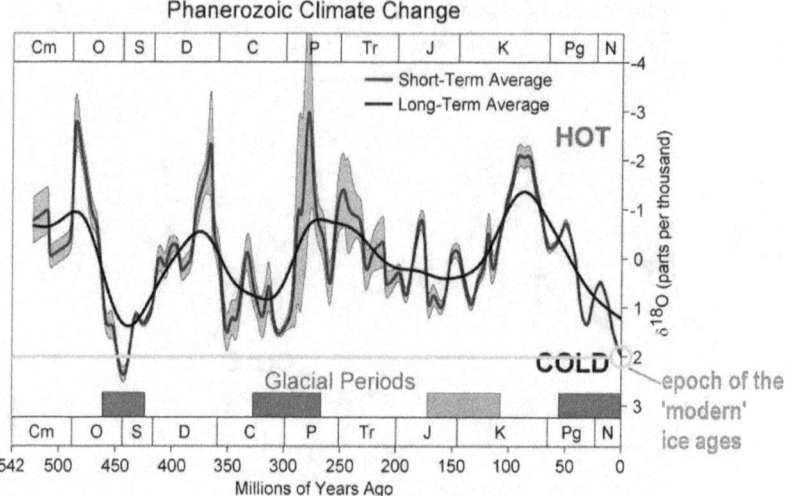

Phanerozoic Climate Change

In these records, we see two deep Ice Age epochs occurring. The first of these lasted about 10 to 20 million years at around 450 million years ago. One of the large mass-extinctions of life occurred in this epoch. Those deep glaciation epochs are rare. The second one is occurring in the present. It started about two to three million years ago.

We know from ice core samples from Antarctica

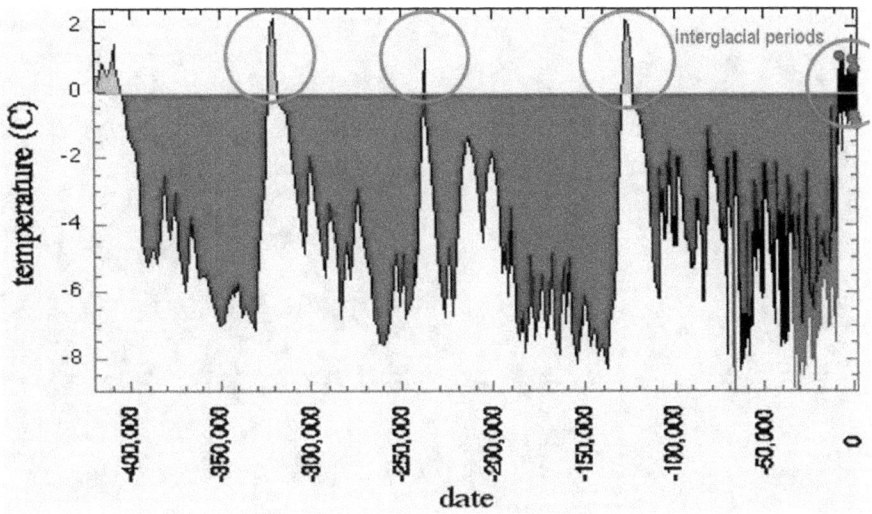

We know from ice core samples from Antarctica that for the last half a million years, the deep glaciation, which may have been caused by the Sun going inactive, were interrupted periodically for a brief span of warm climates, termed the interglacials, of continuous solar activity. We are presently in one of the interglacial warm periods. It has run its course. The plasma density in the solar system is already fading.

Milky Way look-alike
NGC 6744

ESA - Wide Field Imager view - CC BY 3.0

In a galaxy of 400 billion stars, vast networks of plasma streams become entangled, so that the timing of the individual events, at the local level, are best forecast from historic precedents and from present plasma dynamics, such as the fast fading solar wind. Since the Earth's climate is created by the Sun, which is powered by dark matter in the form of plasma streams, reasonable forecasts can be made analyzing the changes in plasma dynamics. One can forecast on this basis that the Sun will likely go inactive in roughly 30 years. While we cannot see the dark matter plasma streams that affect our climate, we can take note of the effects. And if we are wise, we will prepare our world with the type of infrastructures that can enable 7 billion people to live and prosper in the darker and colder world of the coming Ice Age.

Where do the plasma particles come from?

The Implicate Order and Explicate Order

David Bohm
1917 - 1992

One of the questions that has not been addressed in the field of dark-matter science, or plasma physics as it may also be called, is, "where do the plasma particles come from that form the dark matter plasma universe and the atomic elements created by plasma fusion?"
One person has hinted at a possible answer. The man is David Bohm.

Bohm sees cosmic space with an expressed explicate order

The successor

Albert Einstein (1879-1955)   David Bohm (1917-1992)

David Bohm is the man whom Albert Einstein has once referred to as his successor. Bohm sees the so-called empty cosmic space as not being empty as such, but sees it as a sea of apparently disorganized infinite energy that has an underlying implicate order with an expressed explicate order unfolding from it.

The explicate order may be seen like ripples of water

The explicate order may be seen like ripples of water, in terms of latent energy in space.

Expressions of Bohm's explicate order

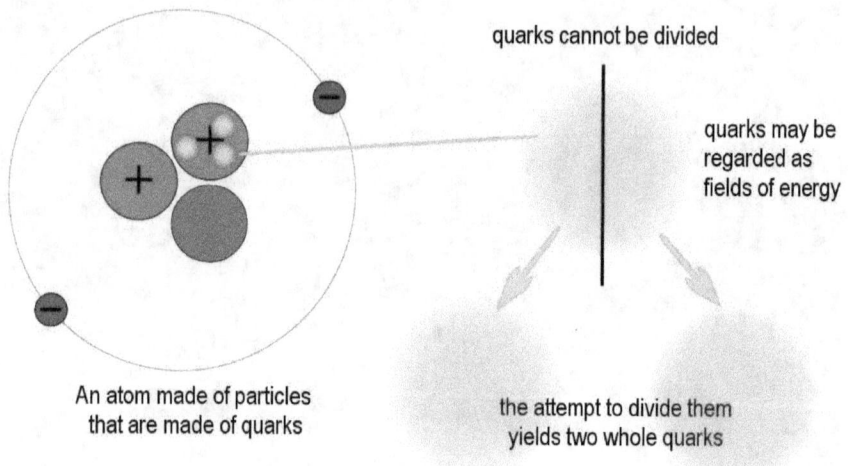

quarks cannot be divided

quarks may be
regarded as
fields of energy

An atom made of particles
that are made of quarks

the attempt to divide them
yields two whole quarks

Since plasma particles, are known in nuclear physics to be
themselves but constructs of points of energy, called quarks, which
are entities that cannot be further divided, it seems reasonable that
one begins to recognize the quarks as but as expressions of Bohm's
explicate order.

In this case, plasma simply exists as an inexhaustible element of
space itself. It may be said in this context that space and plasma and
energy are one, and that from this one the universe unfolds, from
the very large to the very small. While we cannot see the plasma
itself, we can see its dynamic effects.

We can 'see' the start of the Next Ice Age years before it happens

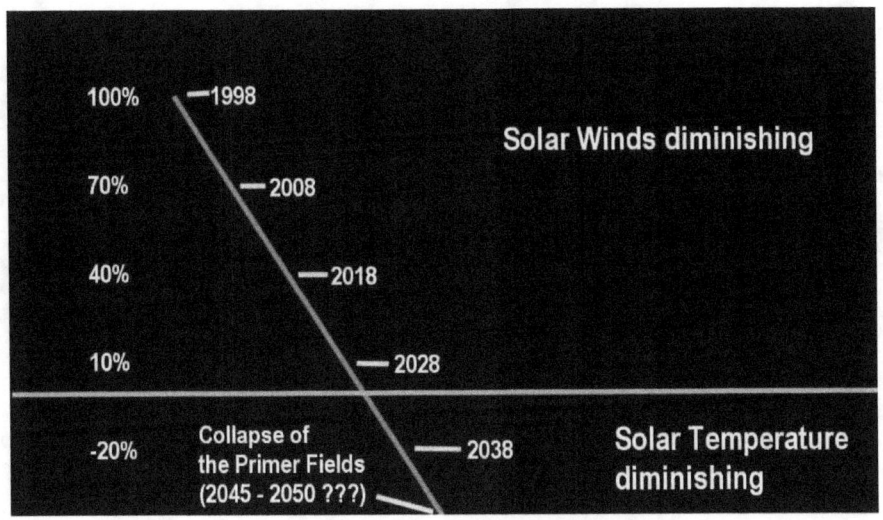

This is also how we can 'see' the start of the Next Ice Age years before it happens. We see the dynamic effects leading towards it. We see the solar winds rapidly diminishing. We see the heliosphere 'shrinking.' We see galactic cosmic-ray flux increasing, together with a lot of secondary effects, such as increasing drought conditions and arctic warming, and so on,

By recognizing the unfolding dynamic trends, it becomes possible to forecast the general timeframe when the Sun goes inactive, by which the Next Ice Age begins.

The capability to look so deep into the invisible

Ice Age of the dimming Sun in 30 years

It becomes further possible thereby, for us to build the infrastructures that we must have in place for us all to be able to continue to live under an inactive Sun. Only humanity, of all the forms of life that we know, has the capability to look so deep into the invisible, and orient its policies in the present accordingly. Of course, as we do this, we thereby discover evermore of our humanity.

False theories inspire fear

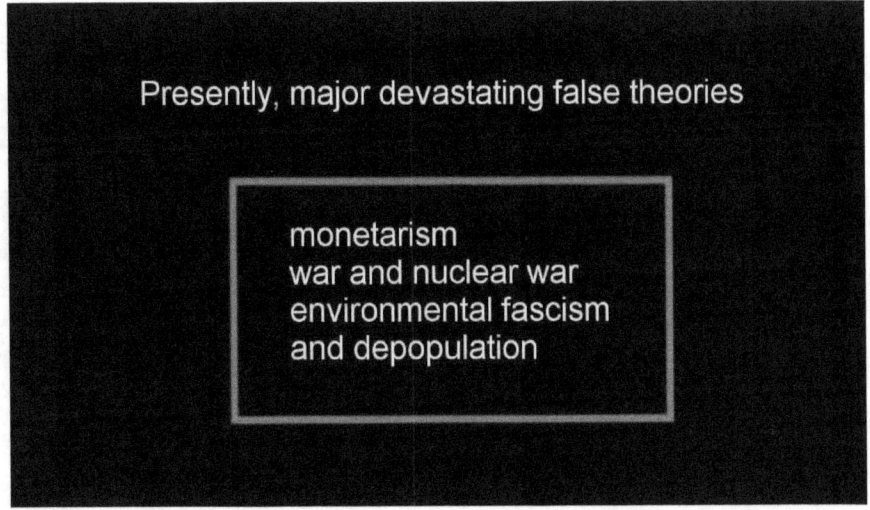

Presently, major devastating false theories

monetarism
war and nuclear war
environmental fascism
and depopulation

As we do this, a lot of long-cherished false theories will likely fall by the wayside, such as orbiting stars, the internally heated sun, global warming, and so on, and of course also the more devastating false theories, such as the theorized need for war, the inevitability of nuclear war and economic collapse and resource depletion, and also the false theory that demands depopulation, and so on. False theories inspire fear. "Fear is the mind killer," said Frank Herbert in Dune. Science and truth open the prison doors of all of the false theories that have chained us, if we strife for the truth to be free.

We stand at the threshold today

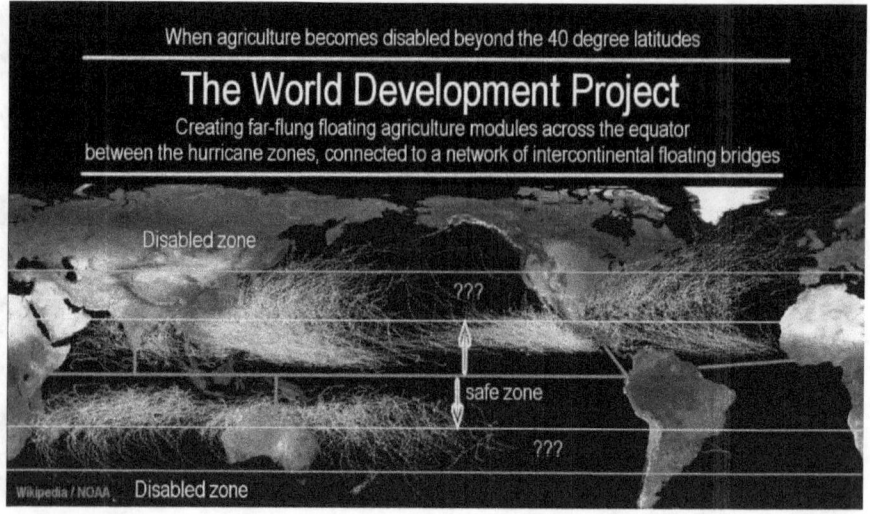

This means that we stand at the threshold today as we explore the nature of dark matter with the mind, which no one can see except with the mind, and prepare our world for Ice Age events that have never occurred in the entire history of civilization, but which we can see occurring 30 years into the future, and this with a certainty as if they would happen today.

That's the dimension of the power of our humanity

Photo by Scott Williams

The Supreme Being

wikipedia

That's the dimension of the power of our humanity. It enables us to step up to higher ground in human living and avoid an otherwise ugly and certain doom. That's how we come of age as a people.